WHY
ROCKS
MAKE YOU
FEEL GOOD

WHY
ROCKS
MAKE YOU
FEEL GOOD

WHERE EARTH SCIENCE PROVES METAPHYSICS

Geology

Metaphysics

CHRISTINE HALL

Table of Contents

Forward

For centuries, humans have looked to the Earth as the source of all life. That life source includes breathable air, clean water, crops to consume and elements to heal.

Today, two battling sciences are merging in the most measurable and spectacular way as we enter an era of exacting measurements. The breakthrough advancements in Quantum Physics and String Theory alone are redefining our understanding of what Energy really is.

Crystals have fascinated humans for eons. Their colors, shapes, density, taste, smell, and frequency have been explored by scholars, scientists, poets and healers for their secrets of the universe. Their powers have been employed by healers and shamans then as well as today in the practices of many current healers. However, the two Sciences; Earth Science and Metaphysical Science have had a disconnect about exactly what does happen when a human life-force connects with Rocks. One camp says, "nothing" and the other camp says "everything".

We are in an unprecedented time in history where the one things all scientists can agree on one thing, EVERYTHING IS ENERGY. And energy can be measured and observed.

The Question:
"WHY Do Rocks Make You Feel Good?"

Introduction

The title of this book represents a big claim. In order to address such a claim, there would need to be a unique set of qualifications line up. Notice the actual question does not ask "do rocks make me feel good," but rather: "WHY do rocks make me feel good?" Which is a much different level of inquiry.

This is a question that I was unable to get any sort of direct answer to which seemed odd but then I realized that the answer lay at the connection point of two semi-incompatible fields of study; Earth Science and Metaphysics. And it is rare to find someone that speaks both languages AND can distill it into real talk that anyone can comprehend easily.

Then upon closer scrutiny of my own resume, I realized I might just be that person that really COULD answer that question. Over the course of my 60 years of experience, experimentation, and education; I am in the most unique position to do so with clarity.

Let's look first at the origin of my claim that I have uncovered a connection between western hard science that can be used to prove claims by ancient healers and shaman's: that rocks; do in fact, have a very real and significant impact on humankind.

Chapter 1

My Background As It Relates To Rocks

My love of rocks started as a child, living with my brilliant grandmother, Ruth Paintin in Carmel, California. For purposes of moving this discussion along, I will refer to all gems, rocks, and minerals as "rocks" unless I'm being more specific.

Ruth held a PHD in Botany that she earned in 1926; which, for the time, a woman to have a doctorate of any kind was unheard of. But earning one in an Earth Science; was extremely rare. She came from stark poverty, as her father barely made a living as a button maker. Work was a collective family effort, as he and his children collected oyster shells on the banks of the Ohio River. She was a remarkable woman, who excelled beyond every boundary presented to her; eventually rising to the respected position of Dean of Women, Chicago University.

Upon her retirement from the field of education, Ruth founded a shop on Cannery Row in Carmel, California, named "Ruth's Gems and Minerals." I often assisted my grandmother in the shop, and at 13 years old; I could sell the heck out of those rocks. I honed my rock selling skills during the mid 1960's, because my grandmother made sure I knew the provenance of the merchandise, in order to further engage the customer in their relationship with rocks. My grandmother Ruth was often consulted by geologists and "rockhounds" as they entered the shop with an unknown specimen. She was the rockhound of all rockhounds. She was their resource as the internet did not yet exist and color photo books were very expensive and rare. Armed with a photographic memory and a willingness to do the deep research; she was as close to a rock "Google" source anyone was going to find.

Most times, upon receiving a rock specimen for consideration, she knew immediately what type it was and would launch into a lecture regarding its characteristics and compounds. I was instructed to observe and absorb...which I gladly obliged. However, sometimes she needed a few days to research it. I would help with the process of figuring out what rock type it was. My first step was to start with a black light that would indicate a fluorescent rock (In fact, a display of fluorescent rocks was my 8th grade science class project). I now know that the black light experiment wasn't really a necessary step; as most fluorescent rocks have other distinctive characteristics. But I was serious about my experiments none-the-less.

But most importantly, she'd see to it that I received the rock "lecture" before it left our scrutiny and was returned to the owner. I've had the privilege of seeing some very beautiful and astonishing rocks in my lifetime! If you want to see some amazing rocks up close, in specimen form (not just a picture); treat yourself to the Smithsonian Museum of Natural History. If a trip there isn't possible today, then check many of them out at this website: https://www.si.edu/spotlight/geogallery.

On days we were not in the shop, we were on the beaches near Big Sur digging for agates and jaspers. The good ones came home with us and went into the rock tumbler, which was located right outside my bedroom window (that's probably where my need for white noise to sleep originated). Thus, also began my jewelry making life. As a child, my jewelry creations consisted of a pretty piece of Jasper that had twine attached by two-part epoxy. And even that wasn't the true beginning of my "rock art." At age 5, I was given a roll of scotch tape and was told to go amuse myself. That resulted in me sandwiching street gravel between two pieces of tape and selling the strips to the neighbors for a penny. It was quite gratifying to see my "creations" taped to a refrigerator door.

My mother, Sue Paintin, was also a "Ruth rock disciple". As children, we had standing orders from the both to keep a keen eye out for geodes, that look like a clump of mud; but will be fabulous when one sees the inside. We faithfully searched, but never found any geode specimens in the back yard of our suburban home in Kansas, But I was forever hopeful… and have never stopped searching!

Sue's passion was cave exploration. No cave was too dark, wet, or bat infested.

These spelunking excursions were where I developed a love of caves, and their intricate formations, in addition, my love of gems, minerals, and rocks. In my lifetime, I have been in and out of caves from one end of this country to the other.

Eventually, it was time to start thinking about a career. But it was the 1970's and my choices as a woman were limited to a nurse (blood-NO), an airline hostess (not pretty enough), or a secretary (can't spell). My father proclaimed I was to go to college just long enough to find a college graduate to marry, and then start having babies. And to be fair, that was the common message to females, so he wasn't saying anything outside the social norm at the time. However, I had other thoughts.

This was an historically interesting time for women. For decades the rhetoric had been that women should avoid degrees in science and math, because women simply do not have the "right kind of brain" for it. Without those disciplines on your resume, a woman could not be competitive in achieving higher paying jobs like engineering, medicine, science, or any other mathematical related field. And even if you could get into a program like geology and receive the degrees; you wouldn't be hired as a woman because it was "man's work" as he must support the wife and child, or "it wasn't lady-like work," or "women don't have the intelligence for the sciences".

During my undergraduate years, laws were being passed. There was a concept being proposed known as "affirmative action." For those unfamiliar with that term, it was first coined by President John F. Kennedy when he signed Executive Order No. 10925, on March 6, 1961. It stated that government contractors had to take "affirmative action" to ensure that all applicants seeking employment would be treated fairly without regard to race, creed, color or national origin.

However, the term "gender" was not added until 1968. That was the social climate for me as I made my decision to become a geologist.

I graduated in 1976 from Coe College in Cedar Rapids, Iowa, with a Fine Arts degree. I chose Coe College because it had the Tri Delta sorority of which my mother was an alumni making me a legacy, (Legacy means shoe-in) and it had no math requirement at the time.

Additionally, Coe also had outstanding internship programs, which I was very fortunate to be awarded two of the best internship opportunities.

My first Internship was historically epic as it required the USDA Forest Service to hire a woman as a Field Scientist, something which had never been done prior to 1975. Affirmative Action legislation was affective as I became the first woman to be hired by the Forest Service in the capacity of scientist; rather than the traditional female role as a secretary. It caused quite the consternation from the concerned wives of my colleagues.

Problem solved in their minds when the only SINGLE man on the Scientific Team of 7 men, who was 12 years my senior, was literally assigned to make sure I didn't temp any of the married men away from their work and wives. Trust me, no temptation on my part, I can assure you. In addition, the Forest Service refused to pay me per the contract agreed upon prior my making the journey from Kansas to Montana. The rational from the director of the Forest Service station offered to me was that I didn't NEED a paycheck, as I had a father that would see to my "keep."

It can be difficult to break through traditional social boundaries in an ever-evolving era. Getting through the door to do the work, and experiencing one-on-one training in the mountains of Glacier National Park and the Flathead National Forest was monumental. I was up against jealous wives and secretaries, resistant colleagues, and dogmatic men who were shocked when I would go ten days into the Bob Marshal Wilderness with seven male scientists carrying a 50-pound pack on my back!

My position was Field Assistant to the Hydrologist, Geologist, Wildlife Biologist, Fisheries Biologist, Landscape Architect, Silviculturist and Botanist. I lived in Kalispell, Montana, and worked out of the USDA office. Most of my work was in the field and every day was a unique experience. One day might have included flying in a small plane over Glacier National Park on a water quality flight. Another day I was conducting wildlife and flower counts and soil analysis. Other days might include helicopter swoops over meadows of elk; or helping avalanches to release in the mountains.

Being the only female within miles did come with some logistical challenges. I literally slept in the panty at the Quintana Field cabin as the six men took the bunks in the main part that also had the only source of heat…the pot belly stove. But that was about the only concession made for my gender and I was treated and taught like any man would have been. I spent hours, days, weeks and months learning at the side of experts in 6 of the Earth Science fields. I hiked around with a 50-pound pack of equipment on my back going in and out of the forest on everything from a truck to a plane to a helicopter to a boat. Nothing was off limits and the learning experience was unparalleled.

I was then offered a second opportunity to expand on what I had learned in the forest, My next stop was Washington DC where I worked on the US Senate Sub-Committee, Materials, Minerals and Fuels. It was headed by Senator Metcalf of Montana and Scoop Jackson was the Chair of the Interior Committee over this sub-committee. It was my job to define two concepts of the House Bill HR 25 as it related to coal strip mining. I was privileged to be able to interview all kinds of experts as it related to the restoration of the land after the mine closes. Specifically, it was my job to define the term, "Return the land to its approximant original contour in arid and simi-arid regions".

As one would guess, it is impossible to return the land to the "approximant original contour" when you've just removed a 30-foot-thick bed of coal, which is the aquifer, which means you just dropped the water table at least 30 feet, which means you wiped out access to ground water downstream to millions of people. Sometimes in arid and simi-arid areas, when that aquafer is removed, there might not be another one for a hundred feet. To complicated matters more, coal mining and Indian water rights are colliding which then brings in all kinds of issues with the violation of the Indian Lands Water Rights Treaties. This made for some very interesting senate hearings when the tribe arrived at the hearings in full tribal dress. Then there is the opposite problem in wet areas where after the removal of a 30-foot coal bed, the mining companies just walked away saying it's restored because now it's a lake. Problem is that it is not a lake as it neither has a freshwater input nor does it drain Hence all the fountains in the lakes in Florida to try and keep the stench and Mosquito hatchery in check.

I was given full access to all government agencies and the top specialists in those agencies as my direct source. Congressional Bills don't like any layers of expert footnotes. I was to go to "The Source" to conduct my interviews. That put me in front of top Earth Scientists in the country at The U.S. Department of Interior, Bureau of Land Management, U.S. Department of Agriculture Forest Service, The U.S. Army Corps of Engineers, EPA, GSS, USGS, and a host of others including mining lobbyists and The Bureau of Indian Affairs. It was unpresented access to a multitude of sciences all rolled into that one little question, what defines the action needed to comply with the letter and or intent of the new law, "Return the land to the approximant original contour".

This experience gave me a close look into mining operations which of course includes oil, coal, rare minerals, gems, iron, gold, etc. Each material requires unique extraction techniques.

At this point I've been field trained to approximately a PHD level, but no one was counting past the BA. Therefore, to be able to merge all this field and life training into a career, I would need to get the correct academics on the resume.

Next stop was Washtenaw Community College because my sad lack of math credits would keep me out of any master's program in

Geology. Remember how I said Coe was on my pick list because it had no math requirement? And believe me, I took advantage of that opportunity by avoiding all math classes. As I mentioned earlier, I received my BA in Fine Arts and managed to avoid all science and math. Time to regroup. I had to get a solid transcript of 4 years of math and science and I had given myself 12 months to get it done. Here is one place I must give credit where credit is due. Washtenaw Community College is the finest teaching institution I have ever known. It might have been bravado or naivety or both, but I went into the college guidance counselor and told her my plan. She said if I can do it, then they can do it.

Mr. McGill was my math teacher, and he literally wrote my math books for me and put me in the math lab where the tutors were. He got me through college level Algebra 1 and 2, Geometry, Trigonometry, Calculus 1, while additionally I was taking Chemistry 1, Biology 1 and 2. I did it all in 9 months and I could not have done that anywhere else.

The next 2 years were spent in the Geo-Building on the Campus of UMKC, University of Missouri in Kansas City. I was the only woman in the program and I had the bathroom all to myself. It's a great program and once again, I had fabulous teachers. I graduated just as the great world oil embargo shut down US oil companies and the only job offers I got were from Juno Alaska and Saudi Arabia.

Then in 1981 I had my first child and travel was no longer an option for me, so I turned my focus from coal, oil and mining to minerals. As I wanted to make a living, I went with a Retail angle and became "mini-Ruth" complete with a Rock Jewelry Store.

11

I said I'd circle around to why all this history is relevant and now it's time to close the circle of understanding. My history and credentials are unusual, especially for a woman, so I got asked by nearly everyone, the same question: "Why do rocks make me feel good?" And in my haste to show off my Geological knowledge, I would launch into everything from Plate Tectonics to Volcanic Action. At which point their eyes would roll to the back of their heads but I'd plow on even though I knew what they wanted. They wanted a Metaphysical explanation that made sense and at the time, I did not believe there was one.

This pattern went on for decades, literally 1000's of people asking me the same thing, "Why do Rocks make me feel good"? People would say, "Oh, that's so cool about you being a trained Geologist, I just love rocks, I pick them up everywhere I go, I still have one from my honeymoon. Why do I love Rocks so much?" And every time I'd evade their real question until one day it occurred to me to take a different approach and consider the possibility that rocks really do make people FEEL good. And not just because they are pretty or interesting or sentimental, but because they physically feel good.

In 1997 I began to attend The Unity Church. This was not so much a deliberate action step on my part but more an accidental action step. My mother was taking on the process of getting sober and she found her way to the AA meetings at the nearby Unity Church. My daughter was also in need of a little less judgmental teen youth group so to support them both, I started to attend the church. I liked the messages and it was Christian based without a dogmatic doctrine and guilt lace messages that many denominations have.

I attended that church for 10 years and met many strong leaders in the New Age Movement. I've met with and spoken to Ram Dass, Bernie Siegel, Andrew Weil, and Jean Huston to name a few. This is a side- benefit of living 3 miles from the International Headquarters of Unity Church. (to be clear, there is a big difference between The Unity Church and the Unification Church whose members are referred to as Moonies).

Although Unity's core message is a very traditional Christian one, it also attracts people who have a strong interest in the world of Spirituality, Heavenly Hosts and Angels. They came with all the tools of that belief system, which include Crystals and rocks. These believers and practitioners would swear they had personally experienced and seen others experience actual physical healing from using crystals in various ways.

I had no problem with understanding the benefits of ingesting certain minerals like Zinc or Potassium. IE: without Potassium, your heart will stop. Those minerals have been well studied and documented as benefitable for bodily function and safely ingested through the lens of western science of pharmaceutical marketability. Most of that research happens in the Universities by the students and those break throughs are owned by the Pharmaceutical Companies.

If there is no monetization of a discovery, it's shelved.

No judgement here, I get the ways of capitalism and free market and I'm all for it. I only point this out to clarify the reason behind the serious lack of Western University science behind Metaphysical Claims.

Without scientific data to back up claims like Rheumatoid Arthritis would be relieved or cured with Fluorite Crystals being held in the one's hand, I was skeptical. These claims come without any real science to back them up, only personal testimonials and I found those to be vague and not held to any actual controlled standards.

Plus, Metaphysics was a new world to me with new concepts most of which left me with big questions. I am seeing the books in the church bookstore that are literally touting the healing practices of rocks for everything from broken hearts (literally and figuratively) to curing cancer. My shock at the audacity to make such claims was largely from not being able to study their scientific data as there wasn't any (as I defined scientific data).

Their data is based on century's old written and oral testimonials and witnessed observations that have been documented for 6000 years. Obviously, this is long before the benefit of electron microscopes, binocular microscopes, X-ray diffraction, microchemical analysis, radioactive detectors, MRI's, CAT-SCANS, etc.

In 2017 my husband and I sold everything and bought a big truck and RV trailer and went off to see the USA. We have gotten to nearly every state and part of our experiences is setting up a tent in small town festivals, large city art shows, psychic fairs and craft shows and offer people the opportunity to participate in my experiment.

This has allowed me the unique opportunity to draw from a huge pool of people to participate with little to no preconceived notions about the whole process as they have never seen anything like it. Basically, the perfect "blind sampling". Some were there just to experience something new; some were there because they really wanted the rock neckless, that I referred to as a Power Shields, but most didn't know why they were there. (Much more on this in chapter 4).

Review:
1. My overall knowledge and experience of rocks has been acquired over a 60-year period. I have extensive field training and formal education.
2. I am well acquainted with the New Age vocabulary and concepts, and I have personally met 1000's of very sane, highly educated, well-traveled, well-read people who have related many personal stories of their experiences with rocks. I've also met and interviewed some of the most respected people in the field of Metaphysics. I became very involved in the Unity Church which allowed me to open my mind about concepts I had previously rejected.
3. I'm a woman in a man's profession which attracted a lot of curiosity in general. Furthermore, women are far more likely to ask a venerable question like "Why do rocks make me feel good?" of a woman than a man. That resulted in an extremely high number of people initiating the question and thereby helping me establish a huge focus group.

Does this qualify me to answer the question? WHY DO ROCKS MAKE YOU FEEL GOOD? YES, with my very specific lifestyles, education, experiences, and general knowledge surrounding these two subjects (Geology and Metaphysics) I do believe it does.

Chapter 2

Basic Geology

There are 4 major branches of earth science, with many other sub-branches. The four majors are geology, oceanography, astronomy and meteorology.

In alphabetic order, this is the short list of some of the sub-studies in Geology (the full list is actually around 150 sub-specialties).

- Biogeology – The study of the interactions between the Earth's biosphere and the lithosphere
- Economic geology – Science concerned with earth materials of economic value
- Engineering geology – Application of geology to engineering practice
- Environmental geology – Science of the practical application of geology in environmental problems.
- Geochemistry – Science that applies chemistry to analyze geological systems
- Geologic modelling – Applied science of creating computerized representations of portions of the Earth's crust
- Geomorphology – The scientific study of landforms and the processes that shape them

- Geophysics – physics of the Earth and its vicinity
- Historical geology – The study of the geological history of Earth
- Hydrogeology – The study of the distribution and movement of groundwater
- Marine geology – The study of the history and structure of the ocean floor
- Mineralogy – Scientific study of minerals and mineralized artifacts
- Mining geology – The extraction of valuable minerals or other geological materials from the Earth
- Paleontology – The scientific study of life prior to roughly 11,700 years ago
- Petroleum geology – The study of the origin, occurrence, movement, accumulation, and exploration of hydrocarbon fuels
- Petrology – The branch of geology that studies the origin, composition, distribution and structure of rocks
- Sedimentology – The study of natural sediments and of the processes by which they are formed
- Stratigraphy – The study of rock layers and their formation
- Structural geology – The science of the description and interpretation of deformation in the Earth's crust
- Volcanology – The study of volcanoes, lava, magma and associated phenomena

The first step to identifying a rock is to determine if it is Igneous, Metaphoric or Sedimentary and often you'll need to see it and understand the field of geology that you are standing on to determine what it is. One of my favorite places in the USA to "see it all" is the Black Hills in South Dakota because it has been everything from an ocean for millions of years, to long ago active volcanos, to earthquakes bringing it all up to the surface again then diving back down, then back up again.

The Earth is in constant motion, from Core to Space. On the surface and sub-surfaces, the crust of the Earth moves in a process called Plate Tectonics and Subduction.

Plate Tectonics is where we need to start to gain a basic understanding of how many rocks are formed. It is a "Theory" and will likely always be considered a theory because you cannot run experiments on this theory due to the scale of the Earth and the fact that we are working in Geological Time which requires millions of years to reproduce an outcome. However, this is the theory that makes everything come together. Keep in mind, Earth Science and Geology are relatively young sciences even though the science's subject is millions of years old. The tools and understanding have only really been around for less than 100 years. Plate tectonic theory had its beginnings in 1915 when Alfred Wegener proposed his theory of "continental drift." Wegener began the conversation and 100 years later, that conversation has been studied and determined to be a viable working theory.

There are three main groups of rocks, Metamorphic, Sedimentary and Igneous. The development of all three of these types of rocks depend on the Plate Tectonics Theory being correct. I'll attempt a super simple explanation of how the Earth is continuing to expand and contract and makes rocks, water and atmosphere (along with everything else that makes up the planet we enjoy).

A good place to start is with the Earth herself as a planet. Let's reference back to our grade school paper-Mache cut away models of the Earth we all had to make (or am I showing my age?). The diagram below is very basic but is all that is needed to continue this discussion into plate tectonics even though there are actually more defined layers than the three main ones, but right now, these three are all we need to know about.

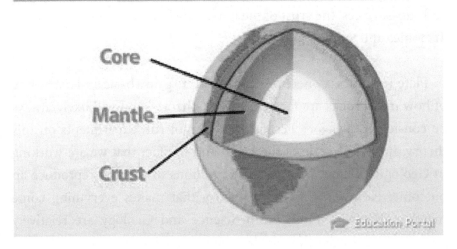

FORMATION OF EARTH'S LAYERS

The center is referred to as the Core. Five fun facts about the Earth's core are:

1. It's Almost the Size of our Moon
2. It's Really Hot!
3. It's comprised of mostly Iron
4. It Spins Faster Than the Surface of the Earth
5. It Creates a Magnetic Field.

Around the Core is the Mantle. Five fun facts about the Mantle are:
1. The Mantle is divided into two sections.
2. There are more elements in the mantle like silicates of iron, sulphides, and magnesium.
3. It cooks at about 5400 degrees Fahrenheit.
4. It is about 1700 feet thick.
5. It is fluid enough to move the Earth's plates around.

And finally, the very most outer layer (if we are not counting our atmosphere) is our beloved Crust. It's the part of the Earth we interact with and live on. There are a zillion Fun Facts about the crust so let's just Geologically understand it as the very thin top layer and it is cold, brittle and filled with rocks.

It is on that brittle breakable crust where we find seven main tears (rips if you will) in the Earth's crust. They are miles long and in addition to those miles of long tears in the crust are hotspots. Hotspots are even deeper holes that find their way from the mantle to the surface, and they are under tremendous pressure, traveling through narrow passages and reaching explosive velocities that allow for them to get to the surface and beyond.

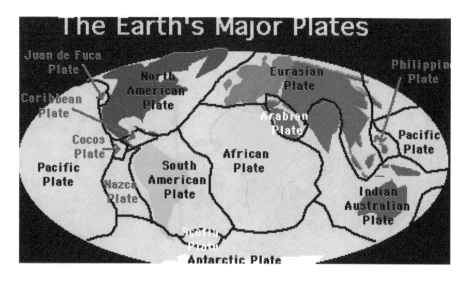

This crude map of the locations of the seven major plates also illustrates the likelihood that the continents did (and currently do) drift apart.

Along those lines is an oozing out of mantle material that creates new crust material when the molten mantle hits the cold water of the ocean and solidifies. As new the new ocean floor material is created, it pushes the older material laterally as the depth pressures of the ocean exceeds the pressure of the expulsion of the mantle material and does not allow the vertical growth. They form giant plates that are floating on top of the plastic-liquid mantle. The more material that is produced by the giant tears in the ocean floor, the more pressure is built up and those plates are literally driven under the continents.

That is a geological process called Subduction in which the oceanic lithosphere (floor) is pushed under the continents and then they are eventually recycled into the Earth's mantle at convergent boundaries. The graphic below illustrates this process.

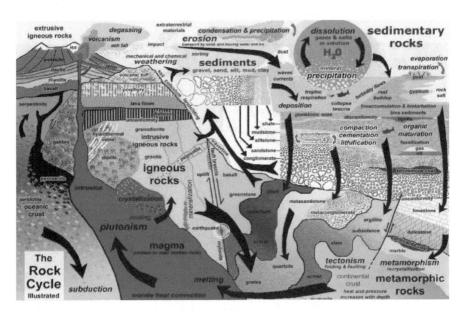

However, when the pressure does exceed the crusts' obstacles, then we have a volcano. When the two events are happening at the same time, you have a plate that moves over a hotspot, which is how Hawaii was built. This is a basic description of the Pacific Plate moving over a high-pressure hot spot that shoots the molten mantle to the surface. At the surface, it "super cools" that mantle material once it hits the air and water. That material is known as Lava and the volcano being built is called a Shield Volcano. The study of volcanos is known as Volcanology (also spelled Vulcanology) and is a whole Science unto itself and if you feel inclined, it's a great one to learn more about.

So now we have Mantle oozing out and cooling and moving plates which are building mountains and breaking apart continents. But wait, there's more. It is all interconnected, from the outer reaches of our atmosphere to the center of the Earth's core. Eons of winds and rains from the atmosphere then break it down again, and it gets pulverized, travels back to the ocean in rivers, gets subducted back into the mantel and the whole cycle begins again.

Understanding the process helps us better understand how to identify and categorize rocks into 3 main categories based on how they are formed. The three basic types of rocks are igneous, metamorphic and sedimentary.

Under each of these categories are even more sub- specialties and this short list serves my purpose in making my point that no one who is an expert on all things "Geological" as it is such a vast subject. Hence, for the sake of simplicity, I'm going to take you on a quick trip through the Mineral Kingdom as rocks are our topic of discussion.

You remember the guessing game where you start with the question, "Is it an Animal or Mineral?".

This is a "chart" I built specifically for the purposes of helping the average person understand the process of identifying rocks. This is not the way rocks are normally identified and cataloged but it is a unique way to relate to them for someone that just wants a basic understanding by comparing it to something we can relate to. This is in no way meant to be an extensive dissertation on Rock identification, but it is meant to quickly give the reader an overview of the rock world by comparing it to a more familiar subject (dogs) and making it all more relatable and understandable.

I'm going to use a dog as the parallel example to help you see the how the vocabulary works. Let's pick Bob my Pug to use as the animal example and Mosaic Jasper as the mineral example.

Bob is defined via the following process:

Kingdom – Animal

Class – Mammal

Order – Carnivore

Family – Canine

Species – Type of dog

Breed –in this example Bob is a European, Short hair, fawn Pug

Name- Bob

Now let's compare that to how we classify and organize our mosaic jasper.

Kingdom – Mineral

Class – Tectosilicate

Order – our Jasper is a Metamorphic rock as opposed to a Sedimentary or Igneous Rock

Family – Calcedony (the mother stone of Jasper which is a microcrystalline of Quartz with many impurities which define the features.

Species - Jasper (Species isn't the actual Geological name,

but it serves as the parallel example place)

Name – Mosaic Jasper

Mosaic Jasper from "raw" to cut polished and set

Since I used Mosaic Jasper as my example in my Rock Kingdom diagram, let's continue looking at it. I have seen hundreds of Jaspers, but this rock is a stand out for me because of how dramatic the coloring is. Although it is commonly found around the world, in the 1970's it was not on my list of minerals to memorize in college mineralogy class, and it was not until it became used in Jewelry that I took note of it.

Although it can be found in the USA mining slag piles, the mining and commercial export from other countries with different economies makes it possible to import this mineral from the large deposits that produce much more dramatic and colorful rocks like in India, Madagascar, Russia, and Uruguay. These imports are dramatic in appearance due to the conditions they formed in. Change the conditions and the rock changes. The first photo is from my personal collection that I acquired from a Colorado slag pile. Albeit it is interesting, it is lacking in the visual interest compared to the sample from India. And why is there a difference if it is the same rock?

To answer this great question, let's go look at how that process works. When the hot semi-liquid molten mantel flows out into the crust but does not reach the surface, this is called an <u>igneous mantle material intrusion</u> ("In" is the clue…. we're referencing everything is happening INside the crust) vs. Extrusion ("ex" is the clue that the mantle material exists the crust).

Those intrusions follow cracks in the crust referred to as faults. We usually hear about faults as they relate to earthquakes. We all have a basic understanding that an earthquake is something where the earth moves, and human disaster is incurred. There are roughly one million earthquakes each year on Earth and the majority of earthquakes only last a few seconds and humans will never feel them. However, some large quakes may last minutes plus after-shocks.

About 90% of all Earthquakes are produced at the plate boundaries where two plates are colliding, spreading apart, or sliding past each other. When large plates move suddenly, they release an incredible amount of energy that is changed into wave movement. On a level of power that we cannot comprehend, that wave will literally travel through rock layers for miles and destroy buildings, roads, towns, and even sink an island. Earthquake waves resemble sound and water waves and can be measured by seismographs which are commonly used around the world to monitor the pressure of these plates as they dance.

Most faults are reverse faults, strike-slip faults, and normal faults, but there are others like oblique faults, and radial faults. Faults are large cracks in the Earth›s surface where parts of the crust move in relation to one another driven by the ever-shifting process plate tectonics. How they collide and move in response to that collision will also determine what the fault will be called.

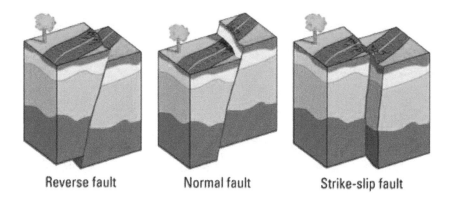

Reverse fault   Normal fault   Strike-slip fault

Once these cracks develop, they create avenues for the molten mantle to enter. Depending on how high up into the crust that molten material goes will determine its cooling time. While it is hanging around waiting to cool, the surrounding material will be incorporated. That incorporation blends in the trace minerals of the rock.

This is not unlike a French baker who uses different ingredients for different end products. When the baker heats and cools pastries at different times and temperatures, we end up with a different product. Ask any good baker what happens with just sugar, water and heat. You can have soft stage candy, caramel or hard stage candy solely based on time and heat. Add additional ingredients and you have toffees. The parallel to Mother Earth's kitchen is the same. She adds a little of this, a little of that, heats it at this temperature for that amount of time and then she has an earthquake to push all her treasures to the surface for us to enjoy.

There is so much more to this process as I'm sure you can appreciate; however, the objective right now is to have a basic understanding of how rocks form.

Chapter 3

Basic Metaphysics

Conversely, on the other side of this equation is the science of Metaphysics which also encompasses many sub-specialties.

The science of Metaphysics is the branch of philosophy that examines the fundamental nature of reality , including the relationship between mind and matter, between substance and attribute, and between potentiality and actuality. (as defined by Geisler, Norman L. "Baker Encyclopedia of Christian Apologetics" page

446. Baker Books, 1999). The word «metaphysics» comes from two Greek words that together literally mean "after or behind or among [the study of] the natural". It has been suggested that the term might have been coined by a first century CE editor who assembled various small selections of Aristotle's works into the treatise we now know by the name *Metaphysics* (*ta meta ta phusika*, 'after the *Physics* ', another of Aristotle's works).

Ref: Cohen, S. Marc. "Aristotle's Metaphysics". *Stanford Encyclopedia of Philosophy*. The Metaphysics Research Lab Center for the Study of Language and Information Stanford University Stanford, CA. Retrieved 14 November 2018

I was quite surprised by how many disciplines are within the umbrella of Metaphysics. There are literally dozens of categories and hundreds of sub-specialties which I found to overlap, contradict, or agree with each other (which not unusual in the world of all scientific disciplines). I have limited the following list of examples of branches of metaphysics that have something to do with rocks and the earth's formation.

Cosmology studies the origin, fundamental structure, nature, and dynamics of all the Cosmos and Physical cosmology is the study of the largest-scale structures and dynamics of the Universe and is concerned with fundamental questions about its formation, evolution, and ultimate fate.

From that starting point, pick your path to study. A few options are actually found in arenas that are not considered Metaphysical. Certainly, one would find the Big Bang Theory heavily studied in many of the western science disciplines like Physics and geology.

Big Bang cosmology (standard) – cosmology based on the Big Bang Theory. The Big Bang is a theoretical explosion from which all matter in the universe is alleged to have originated approximately 13.799 ± 0.021 billion years ago.

Non-standard cosmology – the search for any other explanation of the formation of the cosmos that is NOT the Big Bang Theory.

Plasma cosmology – Theorizes that the dynamics of ionized gases and plasmas and not gravity, play the dominant roles in the formation, development, and evolution of astronomical bodies.

Religious cosmology – A body of beliefs based on historical, mythological, religious, and esoteric literature and traditions of creation and the study of end times.

Abrahamic cosmology – This is based in the Abrahamic religions, including the Bible of Judaism and Christianity.

Buddhist, Hindu, Jain, Taoist cosmology – Defines the formation of the Universe according to their respective scriptures and practices.

For the benefit of making my point here, I only focused on one small branch of Metaphysics and my takeaway was its reliance on stated philosophies that do not adhere to the same scientific method that Western Hard Sciences do. (or do they?) We are talking about oral and written observations that date back thousands of years. Not having the instrumentation to quantify it doesn't negate it.

For the purposes of furthering this discussion, "Earth Science meets Metaphysics", we will stay in the part of the science that deals with rocks.

Metaphysical science certainly predates the science of Plate Tectonics. Plato references crystal healing in his account of Atlantis, relating the stories of the Atlanteans using crystals to read minds and transmit thoughts, however, it was the Ancient Sumerians (c. 4500-2000 BC) that provides the first historical documentation of crystals being used in magic potions and formulas.

Many ancient societies have regarded stones as objects that can be used in the process of healing. It is a practice known as Lapidary Medicine and was practiced in Medieval times, and in the times of the European Pagans and the Chinese (ancient and modern), Native Americans and in tribes around the world. This practice believes that the crystal will connect with the body energy through the body's chakras, or energy points and can realign the body's energy. Throughout history, Shamans and Healers have been working with crystals, rocks and human healing and have passed down their findings for centuries. Unfortunately, certain cultures that used the practice heavily (IE: Native American Indians) have lost nearly all their oral history which means much of their research is lost. However, some of the ancient societies were able to hang onto their research, with the Chinese and the Hindu practitioners having the most information still available in oral and written form.

As I studied many lists, graphics and written descriptors of these medical practices of our ancient ancestors, I see there are differences and discrepancies, (that exists in almost all research science) but there is one thing that is consistent and that is this practice relies on ENERGY from the rock to connect with the ENERGY of the body.

There is a strong Spirituality element to this science that is virtually ignored in the western earth sciences. Although there is a wide field of descriptor words and concepts within this science, the presence of God, Spirit, and other Heavenly Hosts within everything on the planet is a universally accepted concept among believers. Drawing on the spiritual/soul energy that resides in humans is known as the alinement of the chakras. **Chakras** are known as "the energy wheels" and refers to 7 main energy points in the body. When aligned, they correspond to bundles of nerves, major organs, and areas of our energetic body that affect the emotional and physical well-being of the human body.

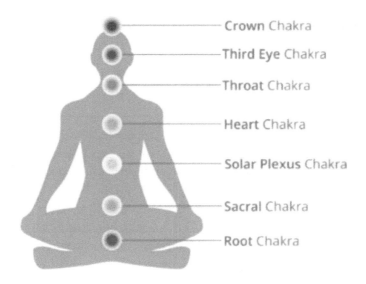

Crown Chakra

Third Eye Chakra

Throat Chakra

Heart Chakra

Solar Plexus Chakra

Sacral Chakra

Root Chakra

Practitioners of using rocks and crystals to reline the chakras are most often referred to as Reiki Masters or Energy Healers. Certification in the practice relies on both the memorization of the chakras and the corresponding rocks and the merging energies. At first, I was put off by the lack of training until I realized that this is a science that relies on implementing what is already "known". Not much different than your doctor giving you an antibiotic for a strep throat without having to go back and re-invent penicillin.

Typically, the person seeking a treatment that uses crystals and rocks for the alinement of their chakras from a reiki master, would lie prone on a table and the practitioner moves energy around the body with the energy of the stone. Depending on what the outcome they are seeking will determine how and what stones are used. The following is a graphic of some of the claims associated with gems and minerals.

	AZURITE / MALACHITE	Skin diseases; anorexia; calms anxiety; lack of discipline; powerful healing force to physical body; emotional release.
	BERYL	Laziness; hiccoughs; swollen glands; eye diseases; bowel cancer.
	BLOODSTONE / HELIOTROPE	Circulation; all purpose healer & cleanser; stomach & bowel pain; purifies bloodstream; bladder, strengthens blood purifying organs.
	CARNELIAN	Grounding; stimulates curiosity & initiative; focuses attention to the present moment; use with citrine on lower 3 chakras; digestion.
	CHALCEDONY	Touchiness; melancholy; fever; gallstones; leukemia; eye problems, stimulates maternal feelings & creativity. Release.
	CHRYSOCOLLA	Emotional balancer & comforter; alleviates fear, guilt & nervous tension; facilitates clairvoyance; arthritis; feminine disorders; eases labor & birth, thought amplifier.

	CHRYSOLITE	Inspiration; prophecy; toxemia; viruses; appendicitis.
	CHRYSOPRASE	Gout; eye problems; alleviates greed, hysteria & selfishness; VD; depression, promotes sexual organ strength.
	CITRINE QUARTZ	Heart, kidney, liver & muscle healer; appendicitis; gangrene; red & white corpuscles; digestive tract; cleanses vibrations in the atmosphere; creativity; helps personal clarity; will bring out problems in the solar plexus & the heart; elimin;:ites selfdestructive tendencies.
	CLEAR QUARTZ	Transmitter & amplifier of healing energy & clarity; balancer, channeler of universal energy & unconditional love; all-purpose healer; programmable
	DIAMOND	All brain diseases, pituitary & pineal glands, draws out toxicity, poison remedy.
	DIOPSITE & ENSTATIATE	Organ rejection; heart, lung & kidney stimulation; self-esteem
	OPAL/ LIGHT	Balances L & R brain hemispheres for neuro disorders; stimulates white corpuscles; helps bring the emotions to mystical experiences; aids abdomen, pituitary & thymus problems.
	PEARL	Eliminates emotional imbalances; helps one master the heart chakra; aids stomach, spleen, and intestinal tract & ulcer problems.

	PERIDOT	Protects against nervousness; helps alleviate spiritual fear; aids in healing hurt feelings & bruised egos; incurs strength & physical vitality; aligns subtle bodies; amplifies other vibrational energies & positive emotional outlook; helps liver & adrenal function.
	PYRITE	Helps purify the bloodstream and upper respiratory tract; upper intestines; digestive aid; nervous exhaustion, grounding.
	QUARTZ/ SOLUTION	Lymphatic cancer & circulatory problems; helps the psychologically inflexible.
	RHODOCHROSITE	Narcolepsy & narcophobia; poor eyesight; extreme emotional trauma; mental breakdown; nightmares & hallucinations; astral body; kidneys; clears solar plexus of blocked energy; unconditional love & forgiveness; evil eye protection; helps one utilize the creative power of the higher energy centers.
	RHODONITE	Inner ear; alleviates anxiety; confusion & mental unrest; promotes calm, self worth, confidence & enhanced sensitivity.
	RHYOLITE	Balances emotions; self worth; enhances capacity to love; aligns emotional & spiritual bodies; stimulates clarity of selfexpression.
	ROSE QUARTZ	Heart chakra opener, love & self-acceptance healer for emotional wounds dissipates anger & tension.

	ROYAL AZEL (SUGALITE or LUVALITE)	L & R hemisphere balance; opens crown chakra; heart expression; increases altruism, visions & general understanding; protects against negative vibrations; helps one gain power to balance the physical body
	RUBY	Heart chakra; balances love & all spiritual endeavors; selfesteem; strengthens neurological tissues around the heart; prevents miscarriages.
	RUTILE	Alleviates blockages within the psyche from childhood pressures.
	SAPPHIRE	Spiritual enlightenment; Inner peace; colic; rheumatism; mental illness; pituitary; metabolic rate of glandular functions; antidepressant; aids psychokinesis, telepathy, clairvoyance & astral projection; personal expression; also for pain.
	SARDONYX	Mental self-control depression anxiety & especially for grief.
	SMITHSONITE	Eases fear of interpersonal relationships; merges astral &. emotional bodies; balances perspective.
	SMOKY QUARTZ	Stimulates Kundallni energy; cleanses & protects the astral field; draws out distortion on all levels; good for hyperactivity & excess energy; grounding.

	SODALITE	Oversensitivity; helps intellectual understanding of a situation; awakens 3rd eye; cleanses the mind.
	SPINEL	Leg conditions, when worn on solar plexus; powerful general healer; detoxification aid
	TIGER'S EYE	Mind focuser; helps purify the blood system of pollution's & toxins; psychic vision; grounding.
	TOPAZ	Balances emotions; calms passions; gout; blood disorders; hemorrhages; increases poor appetite; general tissue regeneration; VD; tuberculosis; reverses aging; spiritual rejuvenationj endocrine system stimulation; releases tension; feelings of Joy.
	TOURMELINE	Olspels fear & negativity & grief; calms nerves; concentration & eloquence improve; genetic disorders, cancer & hormones regulated; raises vibrations; charisma; universal law; tranquil sleep.
	TOURMELINE/ BLACK	Arthritis; dyslexia; syphilis; heart diseases; anxiety; disorientation; raises altruism; deflects negativity; neutralizes distorted energies, i.e. resentment & insecurity.
	TOURMELINE/ RUBELLITE	Creativity; fertility; balances passive or aggressive nature
	TOURMELINE/ WATERMELON	Heart chakra healer; imparts sense of humor to those who need it; balancer; eliminates guilt; nervous system; integration, security & self-oontainment.

40

	TURQUOISE	Master healer; protects against environmental pollutants; strengthens anatomy & guards against all disease; imDroved absorption oF nutrients; tissue regeneration; subtle body alignment & strengthening eye disorders.

There are much more extensive lists but as far as I can determine, these claims are based on, "they say" or "ancient people say". Of course, that is nearly impossible to qualify just based on those references.

As I researched the practices of this science, I noticed a common theme which is the lack of research that really uses scientific measurements or even the organization of applying this to any type of quantitative method. I recently read a passage that went something like, "As you hold the Citrine in your hand, the yellow energy within your body naturally gravitates to it thereby reversing degenerative disease". WHAT IS YELLOW ENERGY?

It is at that moment where the disconnect happens. Where the western scientist walks out, and the eastern scientist meditates on how 6000 years of oral and written historical documentation has just been invalidated.

Chapter 4

The Ion vs. Mother Nature

The stories from the ancient and current shamans and crystal healer didn't convince me on their own merit, I needed more empirical evidence, and thus began my own thought process. The question, "IF this is true, then how?" was not addressed by either side. Both sides made claims: Earth Science says there's no measurable evidence, Metaphysics are not looking deep enough into human physiology. Then it happened, the light bulb went on. If you drill it ALL down to the most fundamental aspect of matter on the Earth, you have the ION.

Then I built from there.

Let me acquaint you to an Ion.

Ions are atoms or molecules where there are a different number of electrons to protons. When there are more electrons than protons, it is negatively charged and referred to as anions. By contrast, when there are more protons than electrons, it is positively charged and referred to as Cations.

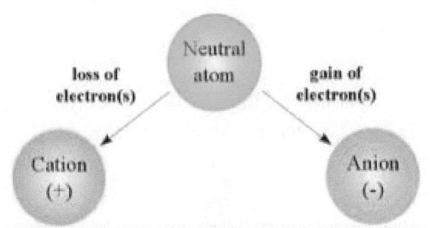

A neutral atom becomes an ion either by losing an electron (cation) or by gaining an electron (anion).

Often, I must ask my audience to ignore semantics and accept the fact that we are negatively charged matter. Examples of other matter that is negatively charged are waterfalls, trees, dogs, animals in general, the ocean, plus ROCKS.

Examples of positively charged matter are most commercial lighting, exhaust from all carbon burning, cell phones, computer screens and even what we refer to as junk food is loaded with added chemicals with positive Ions.

Here is where it gets very interesting. Ions are charges, the most basic aspect of an atom, and just like any positive and negative charge will do when they meet, they will discharge each other. Think of a car battery which holds both a positive and negative charge. If you cross those two cables, then all the Ions, both the positive and negative, discharge each other and the battery becomes inert. We like to refer to that as a dead battery. Bottomline, you present the two opposite charges equally to each other and they will cancel each other out.

Therefore, all is well with your body when you have enough Negative Ions in store to combat the Positive Ions that you encounter.

But what happens when you don't have equal numbers of Negative Ions to combat what you have taken on? Example: You're at the office under all the fluorescent lighting, talking on the cell phone while working on your computer and munching on a bag of corn chips. Then at lunch break you go outside where a dirty truck engine just expelled carbon monoxide all over you. Your waitress has a cold, you cut you finger on your steak knife and come to think about it, you feel like you are coming down with a cold now. You have overdosed your body with Positive Ions which are now known as Free Radicals.

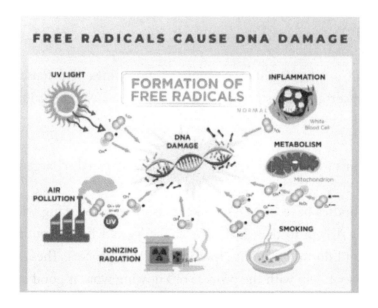

This is where all the Doctors start to get concerned because during the process of cell splitting, the human cell will be looking for building material in the form of Ions. Unfortunately, our cells do not discriminate when it comes to Ion choice, so if there is a Positive Ion there, the cell will use it. That will cause a massive malformation of the cell rendering it no longer effective. Additionally, they are now also coded with two new very bad characteristics; to reproduce geometrically and to NOT die. That is the definition of a cancer cell.

Normal cell
Example of one type of abnormal or cancerous cell

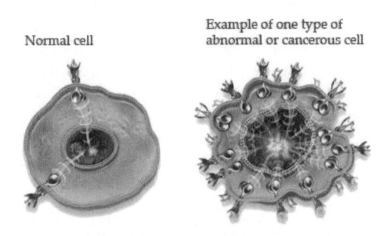

Although our bodies have a couple of more lines of defense against these cancerous cells, stopping them from overwhelming us in the first place is best practice.

And just like discharging the car battery, you can discharge yourself.

Our bodies have one task, to keep us alive. Everything within our bodies is driven to that goal. Our challenge is what we put into our bodies and do to our bodies that impede that process. These days we have a lot of help with the process of knowing what is good and what is bad for us. But before the days of Internet, books or even written language, we had an ally in the battle to keep us alive. Let's call her Mother Nature.

She'll direct us to drink clean water over dirty, or make us afraid of cliff edges, or make us sleep when we're sick.

Sadly, these days we are so busy listening to the "noise" of our lives, we miss the most important voice, Mother Nature. We rely on our sub-conscience mind (Mother Natures' way of communicating) to run most all of our body functions, IE: to make our heart beat and our eyes blink. Mother Nature has a lot of messages for us that we heed without even understanding why and this leads us to this topic of the interaction between the human body and rocks.

One of the most powerful discharging elements of Positive Ions are the Negative Ions from rocks which can meet and render inert large amounts of Free Radicals in our bodies. This is a process that is well understood by Mother Nature which is why, on a primal level, humans have always sought out rocks to decorate our bodies and homes. From the earliest cave drawings, we know that humans have collected, worn, slept on and lived inside of rocks (caves). Since the earliest times, humans have adorned themselves and their homes with rocks with such a passion, it begs the question, "WHY DO ROCKS MAKE US FEEL GOOD?"

My answer to that question is because Mother Nature has conditioned us to repeat successful health practices and reject harmful ones. The science behind the relationship that humans have with rocks comes down to a simple concept, while holding a rock, our bodies don't have to battle off as many positive ions on our own as the rock has neutralized the bad ions for us. And of course, rocks do not stand alone in that "Free Radical clean- up crew". Any naturally splitting running water is discharging pure Negative Ions that our bodies love to soak up. Petting our pets, eating some healthy greens, standing next to an ocean, gives us that opportunity to cleanse ourselves of the free radicals. And when you are doing this cleansing, Mother Nature is using the practice of behavioral modification. She lets you feel that cleansing process, which feels great. Much like the feeling you have once the splinter is out. That feeling that comes over you when you are standing next to a waterfall or river or in a rainstorm is a rewarding feeling of well-being. You are literally feeling yourself neutralizing positive Ions and that feels good. That same feeling is happening when you are holding a rock. And certainly, putting a rock around your neck is much simpler than putting a water fall in your pocket.

Chapter 5

The Experiment

It is my objective to validate both sciences using
The Scientific Method

As a trained Scientist, there is a very specific way to approach such a situation and that is by simply applying the Scientific Method which is defined as "an empirical method of acquiring knowledge that has characterized the development of science which involves careful observation and applying rigorous skepticism about what is observed". Ref: Webster's Dictionary In short it is a 7-step process and you apply each step with due diligence.

1. Question
2. Research
3. Hypothesis
4. Experiment
5. Observations
6. Conclusion
7. Replicate

The Question: "Why do Rocks make me feel good?"

This is the question that has been asked. I wanted to dive much deeper into this question with the addition of the word "WHY". Simply asking the question "Do rocks make people feel good?" was too basic. Stating that rocks "Do" make people feel good didn't go far enough and I knew if I could figure out the "Why", it would naturally validate the "Do".

The Research

The next step was not so easy as I quickly discovered. Seems I was treading on new ground to which little has been written that dives into the core of the question. Too many times the article, book, or blog would either just gloss over the WHY or site unverifiable sources. That is because they had not actually done the experiments themselves.

I found it frustrating that there was no sited research that would back up the big claims that were made about Rocks and Healing. For example, when I researched "The healing properties of Amethyst" I got: "This purple stone is said to be incredibly protective, healing and purifying. It's claimed is that it can help rid the mind of negative thought and bring forth humility, sincerity, and spiritual wisdom. It's also said to help promote sobriety."

Big claims despite its only reference to the original research is "it's said". That certainly would not fly in any grad student's thesis, let alone any western medical professional journal. Any Doctor treating Alcoholism would never prescribe an amethyst necklace as a cure tool. However, many Shamans will and do. What I found was that neither side had enough understanding of the question to find the connection point between the two sciences.

Both Sciences are based on extensive research and solid data. But they do not speak the same language therefore, much is muddled and confusing if you are seeking the answer to that elusive question of WHY do rocks make me feel good? Western answer: "They don't, it's psychosomatic". Eastern Answer, "Because the Moon is in the seventh house" and what does that even mean?

And then I come along, and I am being bombarded with testimonials that rocks do make people feel better and they want to know why. I see it happen, in real life, sitting in front of me, over and over. I listened to 100's of personal stories from people who claim great healings from rocks. It was at that point that I realized that the truth is there, the only thing missing was the connective science between "East and West".

Lack of what I considered credible research material was my biggest challenge when it came to my process of discovery research. You can put this question, "Why do Rocks make me feel good?" into the Google search bar and get a lot of links and blogs and pages of regurgitated "information" that does not state references to actual controlled experiments. Without experiments, it is hard to conduct Observations. Therefore, unlike any other time I have gone down this path, I was actually going to invent my own experiment and do my own observations.

In 2015 my research began in earnest. Certainly, there was a lot of information out there on the basic concepts I have discussed like Ions, Free Radicals, Cancer, Rocks, Minerals, Crystals and even a basic understanding that the human body seeks health and life but there was nothing really connecting it all with anything that even began to qualify as real research. Referencing back to the phrases being used like, "They say" or "It's believed" or "It is said" which doesn't fly in the western world of scientific research.

The definition of research is "the systematic investigation into and study of **materials and sources** to **establish facts** and **reach new conclusions**" ref: Webster's Dictionary.

My research was birthed in a dry field of information. There was plenty of information on each individual subject, (IE: "Metaphysical Healing with Crystals" and "Cancer and Free Radicals" and "Ionic Discharging"). There was even research in the inherent capacity of humans to seek comfort and safety, however, there was very little in the way of making any kind of credible connection between these factors. In other words, there was a serious lack of imperial data to back up many of the claims.

Here's where I begin to do my own research.

At this point, I decided to not only look at where the two sciences diverge but also where they converge. To help you with that process of understanding the two with the least amount of verbiage and redundancy, which is what I have had to slough through, I'll construct this comparison as succinctly as I possibly can and still have leave you with a base knowledge each discipline. Most of the available resources and research was limited to the boundaries of each discipline, therefore, most all my research comes from my own experimentation.

The Hypothesis

The reason WHY people love rocks is they serve as natural detoxifiers. The physical process of the free radical being eliminated produces a measurable rise in people's Dopamine and Serotonin levels. This is the reason WHY you like the feeling of high serotonin levels.

The Experiment

I developed an experiment that allowed me to observe and to be able to quantify those observations. I conducted the experiment for 3 years and I had 100's of happy participants. Here's how it was set up.

I acquired various rocks and minerals in the form of Cabochons. There were roughly 50 specimens to choose from. There was always the large selection of Jaspers, Malachite, Amethyst, Agates, Quartz, Lapis and even some fossils thrown in to make things interesting.

The person I was working with received a beautifully wrapped necklace that I designed just for them, once they had chosen their favorite stone. It began with them sitting at a table across from me that was filled with the stones. They were directed to visually choose their 6 favorite pieces that they liked the look of. Once those choices were made, I had them close their eyes and asked them to keep them closed and just follow my instructions. I placed one stone into each of their outstretched palms and then had them close their fingers up and around them. Then I asked them to open the hand where they felt the strongest reaction (if any). I'd remove both rocks, and I'd set in the special basket the one they resonated with the strongest. We repeated the process until all six stones had between blindly compared to each other and there was a clear winner. Here are my results:

Out of these hundreds of people that had nothing in common, 98% of the participants reported they felt more to the stone than just the surface tension of the rock.

The responses varied and always surprised the participants. Even my most ardent nay-sayers (usually bored husbands) were amazed by the process. On average, the participant reported that of two stones, one in each hand, one or the other stone felt hotter (30%) or felt colder (10 %) or the stone made their arm feel floaty (15%) or the stone gave them tingles (40%) or they felt pressure from the stone (5%). The few participants that felt nothing had a couple common trats that usually translated into a lack of sensation in their hands, or they just willed it to be undetectable. These numbers fall within the acceptable margin of error guidelines.

At the end of the exercise, they were rewarded with their chosen stone that I transformed into a wearable piece of art. Another very interesting thing that happened was that well over 90% of people choose the stone with their eyes closed, that they chose first visually. That I had not expected.

This is research has been conducted from one coast to the other.

In 2017 my husband and I made a radical decision to sell it all, retire, buy a RV and go see the world. That was going to require some extra funds to pull that off, but this dove tailed into what I was already doing in Kansas City. The bonus was that I could find new subjects at new locations if I followed fairs and festivals. That gave me the opportunity to dismiss many factors in the equation. I can absolutely state that this sensitivity to rocks is NOT based on gender, race, religion, age, regionality, or economic standing. That gave me the conclusion that this process works on humans in general.

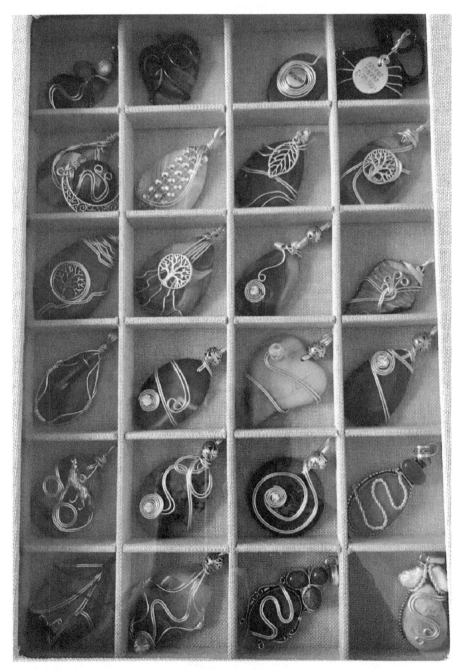

Power Sheilds

The Observation

Starting with the participants general demeaner coming into the experiment, most were healthy, in a good mood, and open minded.

I have broken down my observations down into categories:

1) Participant's enjoyment in the process:
a. Everyone who participated enjoyed the process. Some people had fast responses, and some were slower in terms of response time.
b. The participants took the challenge on with both a real seriousness and a sense of adventure and fun.
c. Each person's time at the table was private and not for public display as I didn't want previous actions dictating the next participant's response. That also protected each participant from any embarrassment by doing this activity.

2) Participant's bonding with final product:
a. Once it was determined which rock they resonated with the most I would try and switch them to a different rock than the rock they chose. I always offered any stone on the table to them, but not once did someone trade. They almost took on an "adoptive parent love" for their chosen baby rock.
b. Nearly 90% of participants also chose their 1st visually favorite rocks with their eyes closed. Initially, I didn't notice this phenomenon as I wasn't looking for it. However, it quickly developed into another interesting thing about this human-rock relationship we have going on. It almost feels like a parlor trick that I can't figure out.
c. These cabochons are unusually large for jewelry as the rock is the forward feature. A very interesting comment I heard a lot was, "I have boxes of jewelry, but I'll never take this one off. The participants'

bonding was stronger to this stone than is normal for a typical jewelry customer.

3) Participant's reaction to the feelings especially if it is their first time:

a. Roughly 75% of people who participated had never felt the energy of a rock before. Although most had heard that there were people who claimed they had felt it, they were skeptical of the veracity of the claims. As I stated earlier, 95% of participants reported feeling something in addition to just the surface of the rock itself. That means that I got to witness the responses of people who had never even heard of such a thing, and they were bowled over that this happened to them. Those were strong responses that I observed over and over. Usually, the process starts with surprise and often ends in either giggles or tears. As they hold the two stones, they are usually narrating what they were feeling. The responses rarely varied from the categories I mentioned earlier, Hot, Cold, Tingle, Floaty, and a small percentage said it felt the rock pushing or putting pressure on their hand.

4) Participant's gratitude for finally being validated for what they have always known

a. The tears usually came at the end when they felt the validation of what they "knew" was true and was taught by Western Earth Science was not true. It's that tear-jerking validation when you know you aren't a freak of nature, and this experiment proves it. It's not a freak of nature, IT IS NATURE!

The Conclusion

Rocks make human bodies feel better by neutralizing harmful Ions in the body. The human psyche is being guided towards our betterment of health constantly and when the subconscious detects a life helpful tool, it will draw us back to it over and over. Hence the reason why we love rocks!

Is it Replicable

Absolutely this replicable. All you need are some rocks of equal size and weight, a table, 2 chairs, and the ability to get people to relax and be in the moment.

Chapter 6

Science Vs. Religion Vs. The Ion

"Science has become the contemporary language of mysticism."
Dr. Joe Dispenza

YES!!!

What a fabulous irony. The war between Religion and Science has raged for centuries.

Just ask Leonardo da Vinci his thoughts on the Roman Catholic Church, or vice-versa.

However, before the nineteenth century, the term "religion" was rarely used. In today's common vocabulary, religion is mostly used as a comparative concept, referring to traits, laws, rituals, and belief systems that could be compared and scientifically studied.

The term "science is also a relatively new term, coming into fashion in the nineteenth century. Prior to this, what we now call "science" fell under the terminology of "natural philosophy" or "experimental philosophy".

Classically defined:
"Religion and science are different aspects of human experience that address different aspects of knowledge. Science is based on empirical evidence and testable explanations, while religion is based on subjective belief and typically involves supernatural forces or entities."

You don't have to look far to see examples of these two worlds in conflict. In fact, the term the *conflict model* was coined and holds that science and religion are in perpetual and principal conflict.

Then something amazing happened. The Veil thinned (a concept I'll address shortly) and revealed way more information (things like Quantum Theory and String Theory, plus we can now measure pretty much anything) and for the first time ever, Science and Religion have suddenly come to understand that they are not only NOT mutually exclusive, but in fact, rely on each other as the other's proof of concept.

As both a Scientist and a deeply Spiritually connected person, this new dawning of enlightenment, for both sides of the aisle, has been thrilling to witness and has deepened my respect for both "Religion AND Science".

Like the parable about the 3 blind men describing an elephant, we are all looking at the same thing but assigning a different explanation due to our orientation. Then we get attached to that limited data and assume it completes the story. We insist that an elephant is a long narrow muscular animal because you can only decipher the trunk. But it's not just a trunk, it's a whole elephant and until the joint vocabulary can be established, you'll never know about the giant toenail the other guy knows about.

Cutting edge for science right now, and "OLD AS TIME for the Shaman, is the concept of "Conciseness". A concept so real and so nebulous at the same moment.

We are acutely aware of how fast humans are evolving at this time in history. Just within my lifetime I have lived through technological, medical, engineering advancements that eclipses everything prior to 60 years ago, collectively. I now live with items and tools that were not even concepts when I was a child.

Jean Houston coined the phrase, "We live in Jump Time now" and expounds on that in her best selling book, **Jump Time: Shaping Your Future in a World of Radical Change.**

As the title would imply, Time is Jumping, not crawling and things are changing, very rapidly.

How is this happening? With each new question, a brand-new modality of exploring that question gets invented. Congressional hearings are going on right now about reverse engineering technology that didn't originate on this planet. This topic isn't so much a debate about did it happen, and more about the future ramifications of such accelerated technology.

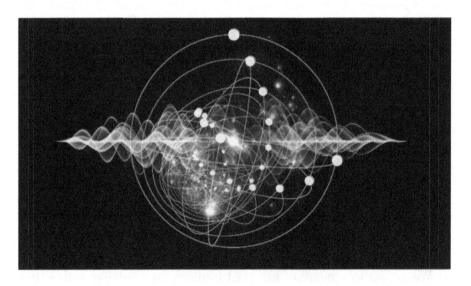

Topping the list of rapid advancements today are:
- Nanotechnology. ...
- Quantum Computing. ...
- Biotechnology. ...
- Materials Science. ...

- Space Exploration. ...
- Genomics. ...
- Renewable Energy Technology. ...
- Robotics. The branch of technology that deals with the design, construction, operation, and application of robots.

Then there's the Rapid Colliders putting the same atom in 2 different places at the same time and now the conversation about dimensions begins.

The number of rabbit holes you can now dive down where Science can now PROVE what Mystics have "known" for thousands of years, is growing. The confluence point is where they both agree, EVERYTHING IS ENERGY.

As Albert Einstein so famously stated (plagiarizing Newton, who plagiarized Aristotle)

"Energy can be neither created nor destroyed but only changed from one form to another." Commonly known as The first Law of Thermodynamics.

And that is where we are going to follow those little Ions that we now know play a key role in both Scientifically and Metaphysically altering human energy.

Quantifiable question, "what is a thought?" What drives it, controls it, manufactures it, confounds it, destroys it?

What particularly fascinates me right now is watching the confluence of the two worlds over the topic known as ENERGY. OBVIOUSLY, it's what this whole discussion is about.

Physis Definition: Energy is defined as the "ability to do work, which is the ability to exert a force causing displacement of an object." Despite this confusing definition, its meaning is very simple: energy is just the force that causes things to move. Energy is divided into two types: potential and kinetic.

Potential energy, stored energy that depends upon the relative position of various parts of a system. A spring has more potential energy when it is compressed or stretched. A steel ball has more potential energy raised above the ground than it has after falling to Earth.May 27, 2024 power derived from the utilization of physical or chemical resources, especially to provide light and heat or to work machines.

Scientists define energy as the ability to do work, or to exert a force that causes an object to move. For example, moving your hand requires energy. Energy can exist in many forms, such as electrical, mechanical, chemical, thermal, or nuclear, and can be transformed from one form to another.

The capacity or power to do work, such as the capacity to move an object (of a given mass) by the application of force. Energy can exist in a variety of forms, such as electrical, mechanical, chemical, thermal, or nuclear, and can be transformed from one form to another.

Metaphysic definition: Internal or inherent power, as of the mind; capacity of acting, or producing an effect.

Notes

Made in the USA
Las Vegas, NV
23 September 2024

95644801R00056